Nuclear Engineering Division 2022

Held at the 2022 AIChE Annual Meeting

Phoenix, Arizona, USA
13-18 November 2022

ISBN: 978-1-7138-7913-8

Printed from e-media with permission by:

Curran Associates, Inc.
57 Morehouse Lane
Red Hook, NY 12571

proceedings.com

Some format issues inherent in the e-media version may also appear in this print version.

AIChE
120 Wall Street, FL 23
New York, NY 10005-4020

Phone: (800) 242-4363
Fax: (203) 775-5177

www.aiche.org

Additional copies of this publication are available from:

Curran Associates, Inc.
57 Morehouse Lane
Red Hook, NY 12571 USA
Phone: 845-758-0400
Fax: 845-758-2633
Email: curran@proceedings.com
Web: www.proceedings.com

TABLE OF CONTENTS

CHEMICAL ENGINEERING APPLICATIONS IN THE NUCLEAR INDUSTRY

THEORY, MODELING AND SIMULATION OF NUCLEAR CHEMICAL PROCESSES II

THEORY, MODELING AND SIMULATION OF NUCLEAR CHEMICAL PROCESSES I

224a Process Challenges and Changes in the Sludge Batch 10 Flowsheet at the Defense Waste Processing Facility

Wesley Woodham, Anthony Howe, Matthew Siegfried

Researchers at the Savannah River National Laboratory (SRNL) have recently completed testing to support the implementation of a chemical processing strategy to treat Sludge Batch (SB) 10 High-Level Waste (HLW) at the Savannah River Site (SRS) in Aiken, SC. Sludge Batch 10 is a paradigm-shifting campaign for HLW treatment at SRS due to the full-scale implementation of several programs. SB 10 will be the first sludge batch campaign undertaken at the SRS that will experience complete support under the nitric-glycolic flowsheet in which the production of flammable hydrogen gas is dramatically decreased by the elimination of formic acid. Second, SB 10 will be the first sludge batch operated from start-to-end with added waste from the Salt Waste Processing Facility (SWPF). Finally, SB 10 will be the first sludge batch processed entirely with the latest antifoamer formulation specified by SRNL. All of these changes reflect a need to understand novel processing challenges that have not until recently been prioritized.

The research presented here will discuss findings from recent SRNL investigations into the effect of the changes discussed above on processing parameters such as flammable gas generation, sludge thickness and processability, reduction/oxidation potential during vitrification, and chemical conversions. Additional discussion will be given to key differences between SB 10 and previous SB campaigns and implications for waste treatment at the SRS going forward.

224b Development and Implementation of the Next Generation Solvent Flowsheet at the Salt Waste Processing Facility

Wesley H. Woodham, Thomas Peters, Seth Hunter

The Salt Waste Processing Facility (SWPF) was recently deployed at the Savannah River Site (SRS) in Aiken, SC. to accelerate the decontamination of salt waste contained within the 51 High-Level Waste (HLW) tanks at SRS. Within the SWPF, the Caustic Side Solvent Extraction (CSSX) process is employed to remove soluble, radioactive cesium from aqueous streams in order to divert the waste streams with the highest radioactivity to the Defense Waste Processing Facility (DWPF) for vitrification. Several potential improvements have been identified in the solvent system currently employed in the CSSX process, leading to the generation of a Next Generation Solvent (NGS) flowsheet to support higher decontamination factors and faster cycle times. Researchers at the Savannah River National Laboratory (SRNL) have performed extensive testing to assess the benefits of this NGS flowsheet. SRNL researchers have also performed the necessary investigations into safety-related aspects of the NGS flowsheet to ensure a successful implementation.

The research presented here will discuss the catalog of SRNL experiments to develop and support the NGS flowsheet as well as recent testing to investigate potential technical and safety-related gaps, such as flammable gas generation, chemical degradation, and differences in physical properties. Additional discussion will be given to the importance of an NGS solvent and the implications of faster salt waste processing across the SRS.

224c Speciation of Aluminum Phases at the Hanford Site

Amy Westesen, Reid Peterson

Treatment of the tank waste at the Hanford Nuclear Reservation in Washington State is one of the largest environmental challenges facing the world today. Disposition of the insoluble solids poses the greatest hurdle to achieving long-term stabilization of the tank waste. The single largest component of the insoluble solids is aluminum. An assessment of the characterization of this waste has identified that the aluminum is roughly split into thirds between sodium aluminate, gibbsite, and boehmite mineral phases. The largest single source of aluminum in the Hanford tank farms is associated with the tanks located in the southwest corner of the site. These tanks contain relatively high concentrations of aluminum such that the mass of insoluble solids present could be reduced by up to 90% through caustic leaching of the materials in these tanks. Chromium is also present and has been suspected of inhibiting aluminum dissolution. An assessment of the impact of chromium (and other metals) on the dissolution of boehmite will be provided.

224d Review of Fire Safety Analyses Conducted at the Savannah River Site

James Laurinat

Fire safety analyses are required for operation of the Department of Energy Savannah River Site facilities, as specified by the site safety basis documents. Analyses required to ensure that flammable mixtures of gases and vapors are not formed are reviewed. Also covered are evaluations of the consequences of postulated vapor deflagrations and detonations and of exposure of radioactive materials to postulated external fires.

Fire prevention measures follow guidelines from the National Fire Protection Association (NFPA). Evaluations of the lower flammability limits for flammable gases and vapors are reviewed, as are methods used to determine the concentration of flammable vapors from evaporation of liquids in open top containers and from spills.

Flammable gas deflagrations and detonations are evaluated to determine if they could breach the containers holding these gases. The maximum deflagration pressure is estimated using the NASA code CEA, which incorporates thermodynamic combustion models. This code also can estimate the pressure for a simple detonation. An in-house code, DPAC, can be used to estimate transient deflagration pressures in unvented and partially vented containers. Peak detonation pressures for normally reflected detonation waves can be estimated by applying a multiplier to the simple detonation pressure. Lastly, detonations require an open gas space with a minimum cell size; rules of thumb to estimate cell size requirements are reviewed.

The consequences of the exposure of radioactive materials to external fires are based on the airborne release fractions (ARFs) and respirable fractions (RFs) for particles released if a container ruptures due to pressurization from heating by the fire. The heating rate for a container exposed to an external fire is evaluated using heat transfer computations, where the container is heated by gas at a specified external temperature. ASTM specifies the temperature transient based on measurements for open pool hydrocarbon fires. A Department of Energy standard gives the ARF as a function of the release pressure and the RF as a function of particle size.

224e Multi-Scale Comparison of Supernatant Tank Waste Hydraulic Performance

Philip Schonewill, Phillip A. Gauglitz, Richard Daniel, Carolyn A. Burns

Execution of the Hanford Site's Direct Feed Low-Activity Waste process requires a pretreatment step prior to vitrification in the Low-Activity Waste (LAW) Facility at the Hanford Waste Treatment and Immobilization Plant (WTP). Tank waste supernatant pretreatment is being performed by the Tank Side Cesium Removal (TSCR) system, which was designed to conduct two separations to prepare LAW for vitrification. The separation steps are (1) dead-end filtration using a Mott Grade 5 sintered stainless-steel filter and (2) ion exchange in packed columns to remove Cs-137 using crystalline silicotitanate. The TSCR separation technologies have been tested extensively at multiple scales with both actual and simulated waste; recently, the TSCR system has also begun treatment operations.

During normal operations in various test systems and the TSCR system itself, the processing flow rate is constant. The dead-end filters and ion exchange columns respond with changes in hydraulic performance, i.e., differential pressure, as solids accumulate in the system or the fluid's physical properties change. The availability of test data collected at multiple scales, including the "full-scale" TSCR, permits examinations of performance to be made based on selected dimensionless parameters. Test data sets will be presented, followed by a comparative examination to assess the viability of making full-scale performance projections from less-than-full-scale test systems, as well as the efficacy of using simulant test data in place of data collected using actual tank waste.

224f Techno-Economic Assessment of Nuclear Production of Hydrogen for Refining Applications in Saudi Arabia

Yousef Alshammari

The Kingdom of Saudi Arabia is planning to construct 16 nuclear power reactors over the next 20 years at a cost of more than $80 billion. It is expected that these reactors will have a capacity of 17 GWe of in order to supply one sixth of the Kingdom's power supply, along with more than 60 GWe of a targeted renewable energy capacity by 2030. As a part of this plan, it would be in the interest of Saudi Arabia to consider the production of hydrogen via nuclear energy, which was already echoed in a recent statement by the Saudi Energy Minister. In fact, the Kingdom is already a major consumer of hydrogen as a feedstock for its oil refineries. Currently, there are 10 oil refineries in the kingdom with a total oil processing capacity of 2.9 million bpd. We estimated that the total demand for hydrogen by all refineries 252 kTonne every year. In this work, the IAEA Hydrogen Economic Evaluation Programme (HEEP) has been used to estimate the LCOE of hydrogen production using four nuclear reactor technology scenarios; Conventional Electrolysis, Conventional Electrolysis with Enlarged Thermal Rating, High Temperature Steam Electrolysis, and the Sulphur Iodine Process. The aim is to evaluate the cost of hydrogen production using different nuclear reactor technologies and how it gets impacted by various economic and technical parameters. It was found the high temperature steam electrolysis using High Temperature Gas Reactors (HTGR) provides the lowest cost for hydrogen production at $\$2.49/kgH_2$ that is comparable to today's production costs from conventional steam reforming of natural gas, Figures 1. It was also found that scaling up conventional electrolysis using boiling water reactors may reduce the cost needed to produce hydrogen from $\$6.2/kgH_2$ to $\$3.97/kgH_2$, Figure 1. The sensitivity of these costs towards various financial parameters including inflation rates, discount rates, tax rates, and decommissioning costs were analysed which provides feasible insight to determine the cost of hydrogen production in various scenarios.

Fig. 1. Cost Estimation of Different Nuclear Hydrogen Production Technologies

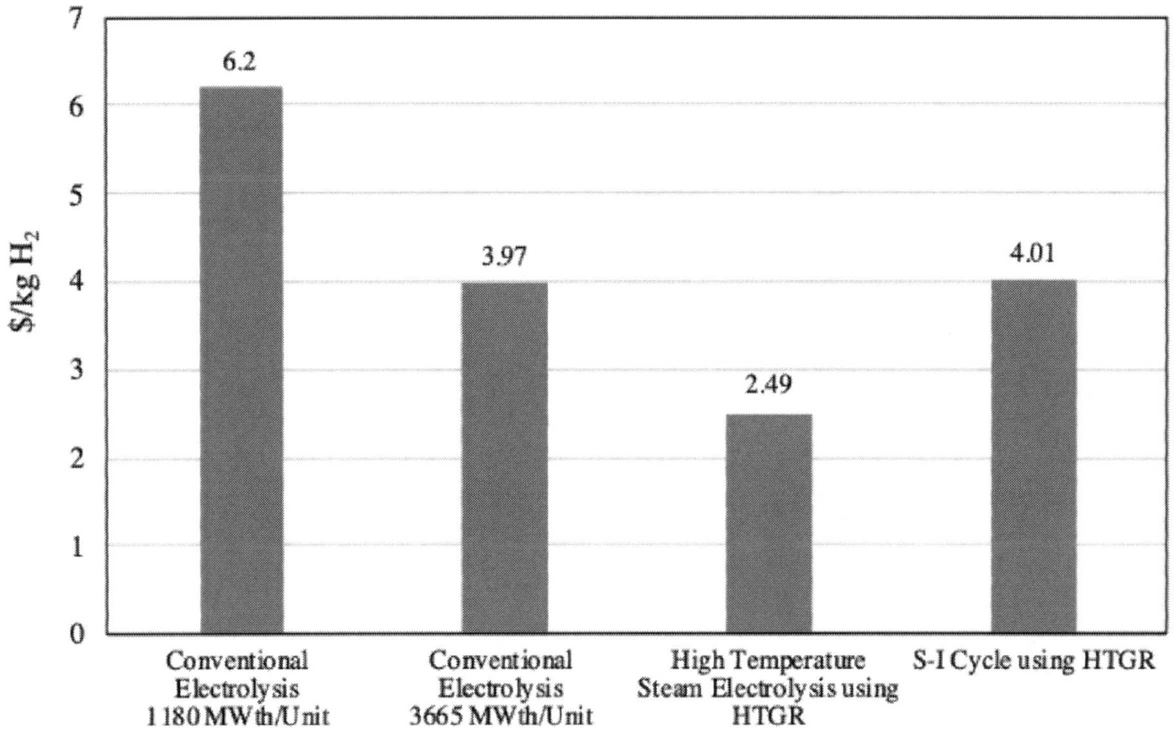

332a Helium Transport and Bubble Growth Near Grain Boundaries in Plasma-Facing Tungsten

Karl Hammond, Dimitrios Maroudas, Brian D. Wirth

We conduct molecular dynamics simulations to study helium behavior near surfaces and grain boundaries in tungsten. Grain boundaries perpendicular to the surface and parallel to the surface are considered. The net attraction of mobile helium to the grain boundary results in a 'denuded zone' within approximately 3.5 nm of the grain boundary plane; helium transport slows considerably once helium clusters encounter a grain boundary. Helium retention is also strongly affected by grain boundaries and grain size: grain boundaries parallel to the plasma-facing direction tend to reduce helium retention due to the attraction of mobile clusters to the grain boundary, which reduces the concentration of helium near the surface, where it is being implanted. Grain boundary planes perpendicular to the surface tend to increase helium retention due to retention of helium on the grain boundary plane. For grain boundaries parallel to the surface, the strong attractive force of the grain boundaries results in essentially no helium transport across the grain boundary plane during at least the first microsecond. It is not known how long such grain boundaries are able to prevent helium from making its way deep into the metal, or whether such a barrier would have any measurable impact on observable phenomena such as the growth of tungsten fuzz.

332b Hierarchical Multiscale Modeling of Surface Morphological Response of Plasma-Facing Tungsten

Dwaipayan Dasgupta, Asanka Weerasinghe, Chao-Shou Chen, Sophie Blondel, Dimitrios Maroudas, Brian D. Wirth

One of the key impediments in harnessing fusion energy for the electrical grid is to develop suitable wall materials for the fusion reactor which can withstand the extreme reaction environment, especially under the conditions expected for plasma-facing components (PFCs) in linear plasma devices and fusion tokamaks. Tungsten is the PFC material for ITER (International Thermonuclear Experimental Reactor) and under the expected typical reactor operating conditions, implantation of helium ions produced in the fusion reaction is responsible for the formation of a fragile fiber-like crystalline nanostructure on the PFC tungsten surface, known as 'fuzz', with detrimental effects on the reactor performance.

Fuzz formation in PFC tungsten is a complex multi-physics phenomenon resulting from the combined effects of driven surface diffusion, subsurface bubble dynamics, bubble bursting and surface crater formation, dislocation loop punching, anisotropies in material properties, changes in material thermophysical properties in the damaged tungsten, and many more. Developing a predictive model capable of capturing the formation and evolution of the fuzz-like complex surface morphology mediated by the underlying dynamical processes that are characterized by disparate spatiotemporal scales has been a major theoretical and computational challenge. Numerous atomistic simulation studies have provided insights into this complex surface nanostructure formation. However, such atomic-scale studies cannot access the spatiotemporal scales of micrometers and hours relevant to fuzz formation.

Here, we present a hierarchical, continuum-scale modeling framework for the surface morphological response of PFC tungsten, capable of accessing the length and time scales characteristic of surface evolution in PFC tungsten. The model accounts for curvature-driven surface diffusion, stress-driven surface transport due to over-pressurized helium (He) bubbles forming in the near-surface region of PFC tungsten during He irradiation, and defect fluxes toward the PFC surface [1]. The model employs a properly parameterized, atomistically-informed equation of state (EOS) for He in over-pressurized bubbles as well as constitutive equations for the mechanical state of the near-surface region of PFC tungsten and is implemented computationally using Fourier spectral methods and semi-implicit front tracking techniques. One highlight of this hierarchical multiscale framework is the calculation of the bubble-matrix (He-W) interfacial free energy per unit area, a key parameter in the EOS for the He density in the bubble [2]. Based on this hierarchical modeling framework, we report results of numerical simulations that predict the onset of fuzz formation in the form of nanotendrils growing from the PFC tungsten surface with an incubation time in agreement with experimental data. We also explore the effects on the surface morphology and growth kinetics of the surface temperature [3], the elastic softening of the near-surface PFC region [4], the helium accumulation kinetics [4], and the formation of nanometer-scale holes on the PFC surface due to helium bubble bursting [5]. The simulation predictions are compared with experimental data and provide a fundamental interpretation to experimental observations.

Furthermore, we have used a cluster-dynamics model to simulate the subsurface helium bubble evolution and bubble bursting and develop a coarse-grained model for He concentration evolution in PFC tungsten, which can be coupled with the surface evolution model. Our cluster-dynamics model, implemented numerically in our code Xolotl, has been parameterized based on large-scale MD simulation results and validated by comparison of its predictions with experimental measurements.

[1] D. Dasgupta, R. D. Kolasinski, R. W. Friddle, L. Du, D. Maroudas, and B. D. Wirth, Nucl. Fusion **59**, 086057 (2019).

[2] K. D. Hammond, D. Maroudas, and B. D. Wirth, Sci. Rep. **10**, 2192 (2020).

[3] D. Dasgupta, D. Maroudas, and B. D. Wirth, Surf. Sci. **698**, 121614 (2020).

[4] C.-S. Chen, D. Dasgupta, A. Weerasinghe, B. D. Wirth, and D. Maroudas, Nucl. Fusion **61**, 016016 (2021).

[5] C.-S. Chen, D. Dasgupta, A. Weerasinghe, B. D. Wirth, and D. Maroudas, J. Appl. Phys. **129**, 193302 (2021).

332c Effect of Surface Vacancy-Adatom Pair Formation on Surface Morphological Response in Plasma-Facing Tungsten

Chao-Shou Chen, Dwaipayan Dasgupta, Asanka Weerasinghe, Karl Hammond, Brian D. Wirth, Dimitrios Maroudas

Tungsten (W) is the chosen plasma-facing component (PFC) material for the divertor in the International Thermonuclear Experimental Reactor (ITER) due to its superior properties, including exceptional thermomechanical properties and low sputtering yield. Numerous experimental studies have established that PFC tungsten suffers severe surface degradation as a result of exposure to high fluxes of helium (He) and extremely high heat loads. Specifically, high fluxes of low-energy helium ions implanted in tungsten within the temperature range from 900 K to 2000 K are responsible for the formation of a "fuzz"-like nanostructure, which consists of fragile nanometer-sized tendrils. The formation of such surface nanostructure has adverse effects on the mechanical behavior and structural response of PFC tungsten as well as on the reactor performance. Moreover, several experimental observations of PFC tungsten surface morphology also have revealed that, during He plasma exposure, aligned stripe patterns may form on the plasma-exposed surface depending on the surface crystallographic orientation, surface temperature, and He plasma exposure conditions. Understanding how such surface patterns form has significant implications for improving the structural and morphological response of PFC materials.

Toward this end, we report here a simulation study of the effect of He-irradiation-induced surface vacancy-adatom pair formation on the surface morphological evolution of PFC tungsten and examine a number of factors that impact such evolution. This surface Frenkel pair equivalent formation has been observed in large-scale molecular-dynamics (MD) simulations of PFC tungsten surface and near-surface evolution upon He implantation at incident energies of He atoms that are not sufficiently high to cause sputtering. Our analysis is based on self-consistent dynamical simulations according to an atomistically-informed, continuum-scale surface evolution model that has been developed following a hierarchical multiscale modeling strategy and can access the spatiotemporal scales of relevance to fuzz formation. The model accounts for PFC surface diffusion driven by the biaxial compressive stress originating from the over-pressurized helium bubbles in a thin nanobubble layer, which forms in the near-surface region of PFC tungsten as a result of He implantation, in conjunction with formation of self-interstitial atoms in tungsten that diffuse toward the surface. The model also accounts for the flux of surface adatoms generated as a result of the surface vacancy-adatom pair formation effect upon He implantation, which contributes to the anisotropic growth of surface nanostructural features due to the different rates of adatom diffusion along and across step edges of islands on the tungsten surface. Using state-of-the-art interatomic potentials, we have computed optimal diffusion pathways along and across island step edges on the W(110) surface and established Ehrlich-Schwoebel (ES) barriers in adatom diffusion across a step edge. Such step-edge diffusion of surface adatoms and ES barriers have been incorporated into our PFC surface evolution model, contributing to surface atomic fluxes, namely, step and terrace diffusive currents, which introduce preferred directions in surface mass transport and give rise to surface features that align to form nanoridge stripe patterns on the PFC surface. We establish that these anisotropic diffusive currents accelerate both nanotendril growth on the PFC surface and the onset of surface nanostructure pattern formation. We also explore systematically the dependence of the PFC surface morphological response on the He ion incident flux and surface temperature, characterize in detail the resulting surface topographies, and compare the predicted surface morphologies with experimental observations. Our simulation predictions for the emerging surface nanostructure patterns under certain plasma exposure conditions are consistent with experimental findings in the literature.

332d Molecular-Dynamics Analysis of the Mechanical Behavior of Plasma-Facing Tungsten

Asanka Weerasinghe, Enrique Martinez, Brian D. Wirth, Dimitrios Maroudas

Investigating the impact of helium (He) ion implantation on the mechanical properties of tungsten is of utmost importance for evaluating tungsten as a plasma-facing component (PFC) in nuclear fusion devices. Toward this end, we carried out a systematic computational analysis of the mechanical behavior of plasma-facing component (PFC) tungsten (W) placing emphasis on how plasma exposure-related defects, namely, voids and helium bubbles, govern the mechanical and structural response to applied straining of the PFC material beyond the elastic deformation regime. Our analysis is based on molecular-dynamics (MD) simulations of uniaxial tensile straining tests (dynamic deformation tests at constant strain rate and temperature) using a properly parameterized machine learning Spectral Neighbor Analysis Potential (SNAP). Our computational models of PFC tungsten consist of regular arrays of He bubbles embedded in the tungsten lattice at different PFC matrix porosities.

We explore the effects on the mechanical behavior and structural response of our model PFC tungsten of a few key parameters such as porosity, bubble size, temperature, and He content. We find that there is a substantial softening of the elastic moduli, following an exponential scaling relation as a function of the porosity and the He atomic content. Beyond the elastic deformation regime, the presence of spherical defects (empty voids and He bubbles) of nanometer-scale size reduces the yield strength of tungsten according to an exponential scaling relation as a function of tungsten matrix porosity and He concentration. Detailed analysis of PFC tungsten structural response near the yield point showed that yielding is initiated by nucleation of twin regions as well as emission of dislocations from bubble surfaces, typically ½⟨111⟩ shear loops, which subsequently glide, grow, and may react to form ⟨100⟩ dislocations. Such twinning and dislocation dynamics facilitate substantial stress relief in the PFC crystal. After this stress relief stage, dislocation annihilation and twin boundary depletion reactions mediate a recovery stage in the PFC matrix, during which the stress increases upon continued applied straining. We also find that, at higher porosities, the higher He bubble density in the PFC crystal has a strong effect on twin boundary depletion during the recovery stage. In general, increase in the dislocation density and decrease in the areal defect density after the initial stress drop upon yielding make dislocation-driven deformation mechanisms dominate the mechanical response of the PFC crystal. Our analysis advances further our fundamental understanding of the mechanical behavior of PFC tungsten and elucidates the role that the most important plasma-related defects play in the mechanical and structural response of the PFC material to applied straining.

332e Elucidating the Mechanisms of Hydrogen Blister Formation in Plasma-Facing Tungsten

Brandon S. Laufer, A. Camila Leiva, Karl Hammond

Hydrogen plasmas are known to form blisters in tungsten. While this a well-known phenomenon, the mechanism of blister formation is not well understood. Evolution of these blisters does not appear to be caused by hydrogen self-trapping or trapping at vacancies, and prior studies have suggested that grain boundaries may play an important role in blister formation. To this end, we used molecular dynamics to study how hydrogen adsorbs onto grain boundaries in tungsten and how adsorbed hydrogen affects polycrystalline tungsten's mechanical properties. We find that hydrogen will collect at grain boundaries and that grain boundaries initially inhibit the transport of hydrogen through the grain boundary plane. For grain boundaries parallel to the plasma-facing surface, however, hydrogen will be 'pushed' through the grain boundary once the entire grain boundary plane is covered with hydrogen. The mechanical properties, in particular the yield stress, will also be altered when the grain boundary is loaded with hydrogen. This may cause delamination of the grain boundaries to occur more readily, leading to blister formation.

332f Analysis of Soret Diffusion of Helium, Hydrogen, and Intrinsic Defects in Tungsten

Dimitrios Maroudas, Dwaipayan Dasgupta, Sophie Blondel, Brian D. Wirth, Enrique Martinez

Plasma-facing components (PFCs) in a nuclear fusion reactor are expected to withstand harsh conditions, with high particle fluxes and extreme heat loads that modify the PFC materials microstructure. These fluxes will create strong gradients of temperature and concentration of diverse species, including He and H atoms and small mobile helium clusters. In addition to these impurity species, neutron particles generated in the fusion reaction will penetrate the PFC material creating intrinsic defects, such as vacancies, self-interstitial atoms (SIAs), and clusters of such point defects. These defects and impurity species will then migrate in the PFC material in the presence of the aforementioned gradients.

In this study, we use nonequilibrium molecular-dynamics (NEMD) simulations to analyze the transport of He, mobile helium clusters, H, and SIAs in the presence of a thermal gradient in tungsten. We find that all the species examined tend to migrate toward the hot regions of the tungsten sample. The resulting species concentration profiles are exponential distributions in the direction of the imposed temperature gradient, rising toward the hot regions of the sample, in agreement with irreversible thermodynamics analysis. For all the species examined, intrinsic point defects and impurities, we find that the resulting species flux is directed opposite to the heat flux, indicating that species transport is governed by a Soret effect, namely, thermal-gradient-driven diffusion, characterized by a negative heat of transport that drives species transport uphill, *i.e.*, from the cooler to the hot regions of the tungsten sample. The findings of our thermal and species transport analysis have been implemented in our cluster-dynamics code, Xolotl, which has been employed to compute temperature and species profiles over spatiotemporal scales representative of plasma-facing tungsten under typical reactor operating conditions, including extreme heat loads at the plasma-facing surface characteristic of plasma instabilities that induce edge localized modes (ELMs). We demonstrate that the steady-state species profiles obtained accounting for the Soret effect vary significantly from those where temperature-gradient-driven transport is not accounted for and discuss the implications of such a Soret effect on the response to plasma exposure of plasma-facing tungsten.

332g Onset of Fuzz Formation in Plasma-Facing Tungsten As a Surface Morphological Instability

Chao-Shou Chen, Dwaipayan Dasgupta, Robert Kolasinski, Brian D. Wirth, Dimitrios Maroudas

Tungsten is the plasma-facing component (PFC) material for the divertor region of the International Thermonuclear Experimental Reactor (ITER) due to its exceptional thermomechanical properties, such as high melting point, mechanical strength, and thermal conductivity, as well as its low sputtering yield, low erosion, and low tritium retention. A large body of experimental evidence has established that PFC tungsten suffers severe surface degradation as a result of exposure to high fluxes of helium (He) and extreme heat loads. Specifically, experimental studies have shown that, under typical operating conditions expected in ITER, He implantation above a threshold of incident ion energy of approximately 35 eV causes formation of a fuzz-like nanostructure on the tungsten surface over the temperature range from 900 K to 2000 K. The formation of such surface nanostructure has adverse effects on the thermomechanical behavior and structural response of PFC tungsten as well as on the reactor performance.

Here, we demonstrate that the onset of formation of the "fuzz" surface nanostructure in PFC tungsten is the outcome of a stress-induced surface morphological instability. Our study is based on numerical simulations and theoretical analysis according to an atomistically informed continuum-scale model for the surface morphological response of PFC tungsten in nuclear fusion devices. Specifically, we show that formation and growth of nanotendrils emanating from the exposed surface of PFC tungsten, a precursor to fuzz formation, is caused by a long-wavelength surface morphological instability triggered by biaxial compressive stress in the PFC near-surface layer due to over-pressurized helium (He) bubbles forming in this near-surface region through implantation of low-energy He ions from the plasma. Using linear stability theory (LST), we predict the onset of surface growth in response to low-amplitude perturbations from a planar PFC surface morphology and calculate the average spacing between nanotendrils growing from the PFC surface. The LST predictions are in excellent agreement with self-consistent dynamical simulations of surface morphological response according to the fully nonlinear surface evolution model starting with random fluctuations from the planar surface morphology that result in nano-scale surface roughness. In addition, we examine the morphological response of the PFC surface to low-amplitude perturbations of very long wavelengths from its planar morphology and interpret fully the simulation results on the basis of a weakly nonlinear tip-splitting instability theory, which predicts a post-instability nanotendril pattern formation with nanotendril separation consistent with the LST predictions regardless of the initial surface perturbation. Finally, we compare our simulation predictions for surface nanotendril growth with experimental measurements of fuzz layer growth on the exposed surface of PFC tungsten. Our simulation results are in very good agreement with the surface growth measurements at the early stages of fuzz growth, further establishing the onset of fuzz formation in PFC tungsten as the outcome of a stress-driven PFC surface morphological instability.

268a Development of Extraction Process Flowsheets for Recovery of Uranium from Used Nuclear Fuel

Candido Pereira, Laura Maggos, Jackie Copple

As part of recent efforts to develop solvent extraction systems for the treatment of spent nuclear fuel, the Argonne Model for Universal Solvent Extraction (AMUSE) was used to design flowsheets to enable recycle of uranium from spent oxide fuels. A complete abstract will be submitted.

268b Nuclear Nonproliferation Stewardship Program - Uranium SandT Center

Jared Johnson

This presentation will provide an overview of a new initiative being executed for DOE/NNSA to maintain US competencies and capabilities in uranium processing expertise. The initiative will focus on maintaining knowledge to design, build, operate, and model unit operations associated with pre- and post-enrichment uranium processing. The initiative will look to establish new collaborations between National Laboratories and academia.

268c Modeling and Simulation for Safeguards at Nuclear Waste Reprocessing Facilities

Anna Taconi, Benjamin Cipiti, Nathan Shoman

Safeguards is paramount nuclear facilities to prevent or detect the diversion of special nuclear material (SNM). Typical safeguards methods for large facilities often rely heavily on destructive analysis (DA) to detect diversion of SNM, but there has been a push to develop safeguards approaches that rely on more heavily on nondestructive analysis (NDA) measurements and process monitoring. The Separations and Safeguards Performance Model (SSPM) can be used to support safeguards by design efforts by modeling process data, key measurement points, and safeguards statistical tests. There are currently multiple versions of the SSPM, including both aqueous and electrochemical reprocessing models, fuel fabrication, enrichment facilities, and Molten Salt Reactors. From the process data, material loss scenarios can be analyzed using simulated measurements and statistical methodologies in the model that detect the material unaccounted for (MUF). The primary statistical method is a Standardized Independent Transform MUF (SITMUF) Page's test, used to detect both abrupt and protracted material loss. The SSPM will be presented along with examples of challenges that have been met through modeling and simulation.

268d Digital Network Twin of the UMass Lowell Research Reactor Facility

Valmor de Almeida, Sukesh K. Aghara, Joseph Boffie

The on-campus UMass Lowell research reactor (UMLRR) is a light-water moderated and cooled, graphite-reflected, heterogeneous, open-pool type research reactor. There is great potential for the UMLRR facility to be used as a test bed for wireless control systems. The facility is fully functional (with availability of current and historical data), well preserved, and consists of key components that can be used for data generation of prospective wireless control components. This work reports on first steps towards developing a digital twin of key components of the facility as described next.

We adopt a divide-and-conquer approach for modeling and simulation of the UMLRR facility wherein significant sub-systems are modeled individually while coupled via an external framework. The computational modules (nodes) in the network are built from stand-alone, reduced order models of the sub-system of interest. Some sub-systems appear repeatedly, either physically or logically, for instance, holdup and sump tanks consist essentially of the same unit operation, therefore a single class module can be programmed for modeling the necessary physicochemical phenomena; similarly the same approach holds for pumps and controllers. In both cases of physical and logical sub-systems, a network can be built for all the components, and a dynamic model assembled to represent a digital network twin of the physical system. A key capability requirement of the framework is to allow fo scalability and re-usability of computational modules. Given the status of modern computing power, extended networks can now be created to various levels of detail; for instance, valves and pipes can be included explicitly in the network. An extended network can be mapped on large scale parallel computing platforms to allow for increasingly modeling fidelity and faster simulation of the evolution of the dynamics of the network.

1) V. F. de Almeida et al., "Cortix: Network Dynamics Simulation," http://cortix.org/ (2018), download at http://cortix.org.
2) T. Azzaoui, K. L. Lyon, and V. F. de Almeida, "Design and Implementation of a Flowsheet Extraction Simulator," Trans. Am. Nuc. Soc., 120, 106–9 (2019).
3) V. F. de Almeida, T. Azzaoui, and K. L. Lyon, "Solvent Extraction Flowsheet-Level Modeling and Simulation with Radiolysis," Proceedings of the Global 2019 International Nuclear Fuel Cycle Conference, ANS, 386–9 (2019).
4) A. Rotker and V. F. de Almeida, "Network Dynamics Simulation of a BWR Plant," Trans. Am. Nuc. Soc., 124, 211–4 (2021).
5) N Kemper, J. Kennedy, and V. F. de Almeida, "Dynamically Coupled Extended Balance of Plant for an SMR," Trans. Am. Nuc. Soc., 124, 215–7 (2021).
6) V. F. de Almeida, "Plant-Level Components for Nuclear Fuel Recycling: Commonality," Letter Report ORNL/LTR-2011/176, Oak Ridge National Laboratory, Tennessee, TN 37831-6181, U.S.A. (Jul. 2011)

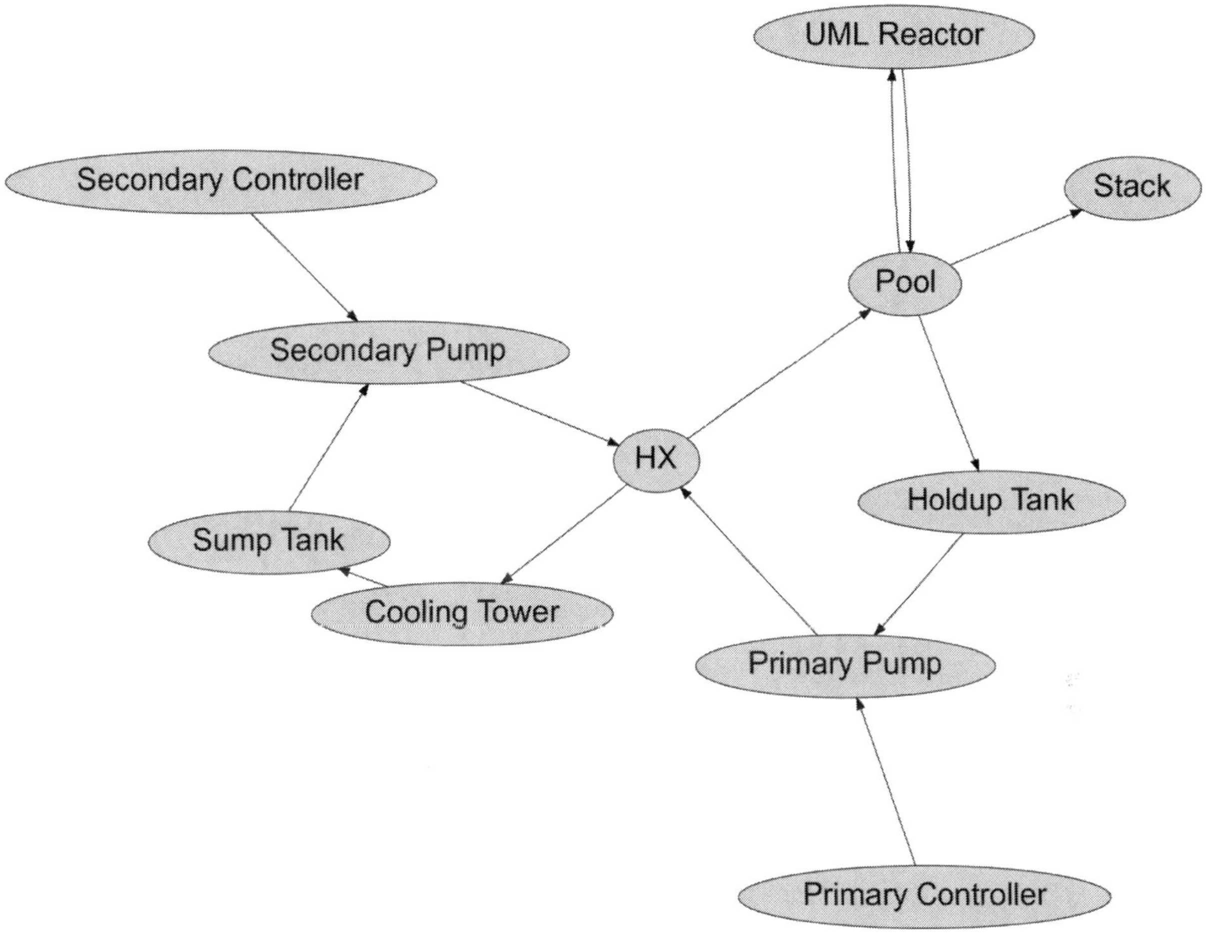
UML Reactor
Secondary Controller
Stack
Pool
Secondary Pump
HX
Holdup Tank
Sump Tank
Cooling Tower
Primary Pump
Primary Controller

268e The Atkau Nuclear Power Plant and Its Relevance to Models of Nuclear Hydrogen

CT Callaway, Nicholas Brown

Nuclear applications to chemical processes such as hydrogen production are attractive of interest to both nuclear and chemical engineers. Hydrogen-based energy systems can greatly offset carbon emissions; however, the high thermal demand of water splitting has posed a significant challenge to the forward progress for large-scale hydrogen production. As the only commercially operated liquid metal reactor coupled to an industrial process, the Aktau Nuclear Power Plant has been the subject of a great deal of research. A thorough review of the plant's lifetime has substantial potential to motivate the development of novel nuclear reactor design, and advance nuclear hydrogen production.

The Aktau nuclear power plant was operated on the coast of the Caspian Sea in what is now Kazakhstan from 1973 until 1999. Mainly used for electrical production and Multi-Effect Distillation desalination, the blueprint provided by Aktau's cross-plant operation including boundary conditions, design-basis incidents, irradiated materials, and equipment safety for the main thermal process can better inform current models for thermal coupling. For example, coupling modeled hydrogen production processes to advanced reactors, such as sulfur-based thermochemical water splitting to a molten salt reactor.

This presentation will focus on compiling lessons learned from the Aktau plant, and their applications. Covering in detail the information gained from the desalination processes and its performance in relation to the BN-350 sodium cooled fast reactor. These lessons were applied to models of a hybrid sulfur hydrogen production process verified by flowsheets made in ASPEN Plus and run in conjunction with a nuclear reactor model.

The coupling of the hybrid sulfur cycle to advanced reactors and its relation to other nuclear powered chemical processes will be presented. The operational details of this modeled coupling informs current generation reactor models such as the molten chloride fast reactor. Finally, the presentation will demonstrate the model's future use in the development of molten salt reactors, and the future potential for MSR driven chemical processes.

268f Properties of Tri-Butyl-Phosphate from Polarizable Force Field MD Simulations

Faranak Hatami, Valmor de Almeida

Tri-n-butyl phosphate (TBP) extractant has been used since the late 1950s to recover metals from used nuclear fuel and other industrial applications in solvent extraction processes. Molecular dynamics (MD) simulations were carried out to calculate thermodynamic properties, transport properties and microscopic structure information of TBP liquid at ambient conditions.
Over the years, several force fields (FF) have been developed (polarizable and non-polarizable) to predict the properties
of TBP, and a number of articles have been published using non-polarizable classical molecular dynamics for studying various aspects of chemical complexation in solvent extraction containing TBP. However, the current capability of atomistic simulation to predict properties are relatively poor. In our previous study, the performance of the OPLS polarizable FF and two sets of non-polarizable force field models, namely OPLS-DFT and OPLS-MNDO for TBP in the liquid state have been examined to calculate properties of pure TBP. Following our previous work, we have obtained excellent results for predicting the mass density of TBP (as expected), for example, the mean value obtained is 0.97023 g/cm3 (sigma=0.011) at ambient conditions which gives 0.08% error when compared to the experimental value.
But we find that our MD simulations using both types of force fields cannot satisfactorily reproduce transport properties such as viscosity and self-diffusion coefficient, specifically, results tend to under-estimate these properties; some of the current results such as mass density and dynamic viscosity are shown in the following figures.

As these results may not be appropriate for reproducing the properties of TBP, further work is needed to explore the possibility of performing improved calibrated polarizable FF. We continue to investigate the possibility of using MD to compute properties with results closer to experimental values. MD simulations on liquid TBP were performed on systems in cubic cells and equilibrated in an isobaric-isothermal (N-P-T) ensemble at 1 bar and 298.15 K for 2 ns. We have been using the Drude package in the LAMMPS MD code to implement atomic polarization. In this work, we are specifically studying the performance of OPLS polarizable FF for TBP to compute: mass density, heat of vaporization, dynamic viscosity, self-diffusion coefficient, thermal expansion coefficient, and dipole moment. This investigation also constitute a comparison between polarizable and non-polarizable FF models, where TBP has both polar and non-polar groups of atoms. It is expected that this work can provide recommendations on force fields for TBP to be used in future additional studies.

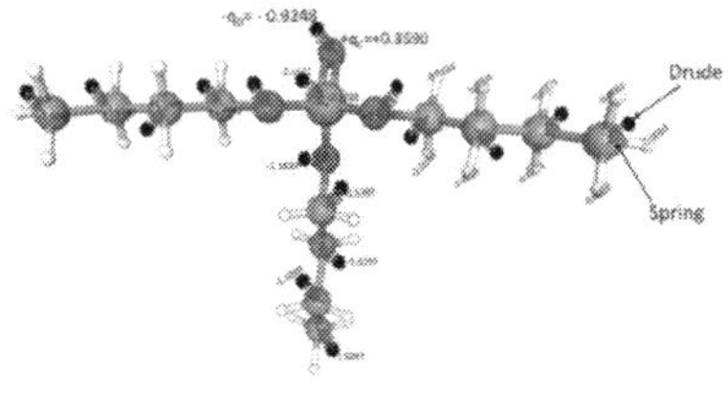

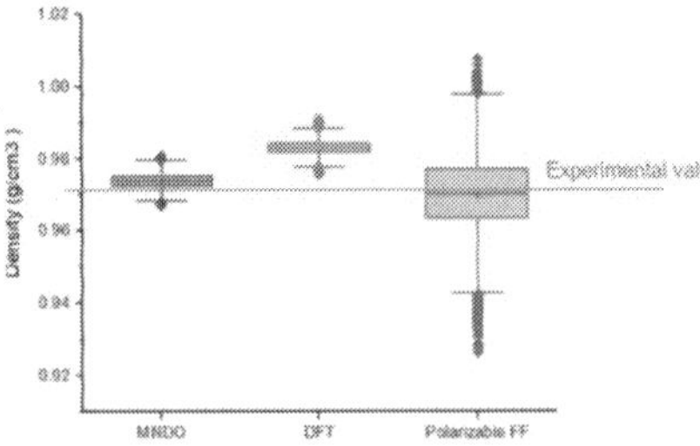

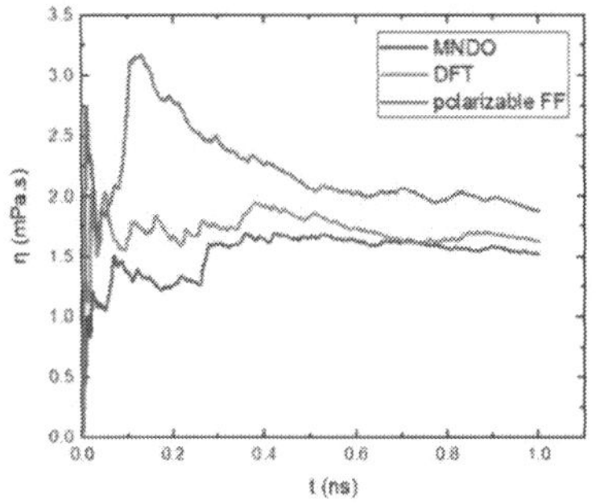

268g Radioisotope Target Dissolution Modeling and Simulation

Michael Dion, Valmor de Almeida, David Abrecht, Andrew Conant, Ken Dayman, Riley Hunley, Richard Reed, Charles Weber

The Radiochemical Engineering Development Center (REDC) at the Oak Ridge National Laboratory (ORNL) develops target material and processes the irradiated targets for producing radioisotopes of interest to the scientific community. The targets are irradiated at the High Flux Isotope Reactor at ORNL. This presentation describes system-level modeling and simulation of the dissolvers at the REDC (15 L and 170 L) for carrying out the caustic dissolution of the target cladding, followed by the acidic dissolution of the remaining oxide pellets. System-level of theory is used to describe the time-dependent chemical kinetics of dissolution of fully mixed systems including the solid, liquid and vapor phases. The modeling and simulation approach is similar to the dissolution of aluminum-cladded metal fuel in past research[1].

The dissolution of irradiated targets liberates volatile isotopes from the target matrix. Parallel research efforts are modeling the subsequent effluent propagation through a physics-based mass balance model tool. Results of effluent transport and temporal radionuclide profiles based on calculated target inventories and process models will also be presented.

1) Key, B., Sadasivan, P., Fallgren, A., Demuth, S., Aleman, S., de Almeida, V.. Chiswell, S., Hamm, L., Tingey, J., 2017 Facility Modeling Capability Demonstration Summary Report, LA-UR-17-20737, https://doi.org/10.2172/1342830.

AUTHOR INDEX